Fires

Elaine Landau

Franklin Watts
A Division of Grolier Publishing
New York • London • Hong Kong • Sydney
Danbury, Connecticut

For Alexander Garmizo

Note to readers: Definitions for words in **bold** can be found in the Glossary at the back of this book.

Photographs ©: AP/Wide World Photos: 33, 36, 37, 39, 40, 46 left, 47 right, 49, 51; Archive Photos: 3 bottom, 9, 13, 18, 27; Art Resource, NY: cover (Scala); Brown Brothers: 17, 23, 24, 26; Chicago Historical Society: 14; Corbis-Bettmann: 44 (Jack G. Moebes), 3 top, 4, 6, 7, 29, 30, 34, 38, 41, 42, 46 right, 47 left, 48 (UPI), 22 (Baldwin H. Ward), 45; Stock Montage, Inc.: 8, 11, 15.
Cover: A painting of the Great Chicago Fire published by Currier & Ives

Visit Franklin Watts on the Internet at:
http://publishing.grolier.com

Library of Congress Cataloging-in-Publication Data

Landau, Elaine
    Fires / by Elaine Landau
       p.   cm.— (Watts Library)
    Includes bibliographical references and index.
    Summary: Examines the causes and events of some of the deadliest fires in the world, how they were fought, the effects on those involved, and what steps were taken to prevent similar disasters from occurring again.
    ISBN: 0-531-20343-3 (lib. bdg.)   0-531-16423-3 (pbk.)
    1. Fires—United States—Juvenile literature. 2. Fire prevention— Juvenile literature. [1. Fires. 2. Fire prevention.] I. Title. II. Series
TH9448.L36 1999
363.37'9—dc21

98-49175
CIP
AC

©1999 Elaine Landau
All rights reserved. Published simultaneously in Canada.
Printed in the United States of America.
1 2 3 4 5 6 7 8 9 10 R 08 07 06 05 04 03 02 01 00 99

Contents

Prologue
The Nature of Fire 5

Chapter One
The Great Chicago Fire 7

Chapter Two
The Triangle Shirtwaist Fire 19

Chapter Three
The Cocoanut Grove Nightclub Fire 31

Chapter Four
The Hartford Circus Fire 43

52 **Timeline**

54 **Glossary**

57 **To Find Out More**

60 **A Note on Sources**

61 **Index**

Flames consume the remains of the canvas circus tent at the Hartford Circus fire on July 6, 1944.

The Nature of Fire

Uncontrolled spreading fires have often resulted in tragedy. When these fiery outbreaks occur in nature, we can only pit our firefighters and their equipment against the blaze. But what about fires that are either accidentally or purposely started by humans? How can we deal with incidents in which carelessness, greed, or criminal behavior were factors?

This book examines a few such tragic fires in which hundreds of people were affected. It explores how and why these

blazes began, and how they were fought. Often, both heroes and villains emerge in the stories, and at times the government's role is called into question. More importantly, the text looks at steps that can be taken to prevent the situation from occurring again.

The charred remains of the city of Chicago, 1871

Chapter One

The Great Chicago Fire

In the fall of 1871, the city of Chicago was destined for a massive fire. Within the previous three decades, the city had grown rapidly and many of the new buildings had been constructed quickly of inexpensive wood. Throughout much of the city, stores, schools, and houses were made from wood. Newly built homes with brick or stone structures had wooden top floors, or roofs with wooden shingles.

People flee across one of Chicago's many wooden bridges to escape the flames.

Hazardous Conditions

In the poor neighborhoods of the city, thousands of people lived in **flammable** wood tenements that were connected to one another by a wooden fence. One Chicago newspaper reporter described the setting: "The land was thickly studded with one-story frame dwellings, cow stables, pigsties, corn-cribs, sheds innumerable, every wretched building within four feet of its neighbor and everything of wood." Chicago also had a wood-based transportation system. The city was tied together through a network of wooden sidewalks and twenty-four wooden bridges along the Chicago River.

The summer before the fire, Chicago had suffered a near drought. From early July to the beginning of October 1871, the city received only a fraction of its usual rainfall. Dry, fallen tree leaves lay in piles on the streets, and the wood in the city had grown dry.

Lloyds of London

Lloyds of London, a famous British insurance company, felt Chicago had become too much of a high risk to insure against fire. The company flatly refused to insure any new Chicago properties and, wherever possible, attempted to cancel existing policies.

Chicago fire chief Robert Williams was well aware of the fire risk to his city. Many times he had asked the city council for additional funding for firefighters and equipment, but the council believed that Williams's team of less than two hundred firefighters was sufficient. The councilmen knew that in the event of a sizable fire, firefighters from throughout the city would be summoned. But they failed to realize how long it would take the firefighters to respond.

The city council had also turned down Fire Chief Williams's request for a fireboat, which would prove to be a grave error. All of Chicago's bridges were made of wood, and a few lumber houses lined both sides of the Chicago River.

This engraving depicts the large number of wooden warehouses and lumber houses that surrounded the city of Chicago.

O'Leary's Barn

On October 8, 1871, many of Chicago's firefighters were still exhausted from battling a sixteen hour-fire the night before. Unfortunately, they weren't able to rest very long. That evening, the fire alarm sounded again, alerting them to the outbreak of another blaze. One resident described the city's mood, "The churches were just dismissing their devout worshippers after evening service, when the fire-bells rang their loud alarm. The evening before, a fire had raged of unparalleled violence, and the embers still glared in the darkness. . . . [Now] many hastened to the scene of the fire, fearing that the high wind might imperil even larger districts of the city . . ."

The blaze began on the corner of DeKoven and Jefferson streets in the O'Leary's barn—a small wood structure filled with hay to feed the few cows and horses kept there. It remains unclear how the barn burst into flames. Many believe that Katie O'Leary, the owner of the barn, went to either milk one of the cows or tend to a sick calf, when the animal accidentally kicked over the kerosene lamp she had with her. Years later when Katie O'Leary's son became a well-known Chicago politician and gambler, he claimed the fire was started by some wayward youths who crept into the barn that night to smoke cigarettes. Another story suggests that the fire was a result of chemicals deposited in the soil by a comet passing over the area thousands of years ago. The comet's residue was supposedly very **combustible** in the area of O'Leary's barn.

However, recent research suggests that Katie O'Leary, the person most often blamed, was actually innocent. According to Richard F. Bales, an attorney with the Chicago Title Insurance Company, the fire is now believed to have been set by a one-legged horse-cart driver. A neighbor of the O'Learys known as Daniel (Peg Leg) Sullivan, was in the O'Leary barn either lighting a lantern or smoking a pipe, which he acciden-

Katie O'Leary milks the cow that allegedly caused the fire in Chicago when it kicked over her kerosene lamp.

Fire Insurance

An owner of a building and its contents may encounter financial loss if there's a fire in the building. However, an owner may reduce or eliminate this risk by purchasing fire insurance. For protection against loss from a fire, a person pays an amount of money based on the value of the property they own. If a fire strikes, an insured individual receives payment on the amount of the loss.

tally dropped. Bales came to his conclusion after reading old records and reconstructing the O'Leary's neighborhood. He noticed that during the official **inquiry,** Sullivan claimed to have been in front of another neighbor's home when he spotted flames coming from O'Leary's barn. Yet property records indicate that Sullivan's view would have been blocked by at least one home and an 8-foot (2.4-meters) fence.

Human Error

Regardless of how it started, the fire that began in O'Leary's barn that evening came to be known as the Great Chicago Fire. Besides the abundance of wooden structures, the dry weather, and insufficient firefighters, there was another reason for the blaze's extremely rapid spread. Unfortunately the "fire watch" (the person whose job it was to look for outbreaks of fire) in the courthouse tower, reported the wrong location. Having estimated that the blaze was more than a mile away from the actual site, city firefighters headed for the wrong address.

Fire Alarm!

During the 1600s, firefighters were called to a fire by a "rattle-watch"—a loud-sounding wooden rattle, accompanied by loud shouts from the fire patrol on duty. Fortunately, in 1840, Dr. W. F. Channing invented the fire-alarm telegraph system, which was used until the early 1900s. By 1970, fire departments throughout the United States were using a mobile narrow-band FM radio, with more than eighty-two channels for communication.

Once he realized his error, the fire watch contacted the **dispatcher,** urging him to send out a second alarm. But the dispatcher refused, claiming that another alarm would only confuse the firefighters. As a result, by the time most of the firefighters reached O'Leary's barn, several city blocks were already in flames.

In an effort to save the city, firefighters work to extinguish the flames.

Fear Spreads

The fire continued to spread as wind-borne sparks quickly **ignited** other parts of the city. One landed on the ornate steeple of St. Paul's Cathedral, setting the church on fire. Propelled by the evening's strong wind, the fire advanced

St. Paul's Cathedral stands among the ruins of Chicago.

northward before reaching a lumber yard. Within an hour, three massive fires consumed Chicago. The fire was so intense and widespread that thousands of people waded into Lake Michigan and crouched down in the water with their backs facing the city to avoid being burned to death.

Firefighters battled the flames as the night grew darker. But Fire Chief Williams, along with many others, had begun to see the hopelessness of the situation. Before long, panic spread through the city. Families who had gone to sleep thinking the fire would be contained woke to the sound of fire bells and urgent warnings to leave their homes. A Chicago minister, Reverend E. J. Goodspeed, described the scene:

"The angry bell tolled out, and in a moment the bridges were choked with a roaring, struggling crowd. . . . The people were mad. . . . They stumbled over broken furniture and fell, and were trampled under foot. . . . They surged together backwards and forwards in the narrow streets, cursing, threatening, imploring, fighting to get free."

Reinforcements

Desperate for help, Chicago mayor Roswell B. Mason telegraphed other cities to request additional firefighters and equipment. In response, **reinforcements** from Cincinnati, St. Louis, and Milwaukee headed for Chicago. Meanwhile additional problems arose. People tried to save large, heavy posses-

Many of Chicago's citizens attempt to rescue items from the burning Crosby Opera House.

sions by throwing them out of windows. But in many cases the objects hit people in the streets, injuring or even killing them.

When the flames reached the courthouse, the police had to let the prisoners out of jail. They managed to do so just before the building collapsed. These convicts might have been the only few people to actually benefit from the fire.

Numerous buildings in Chicago's business district collapsed as well. The staff at the Chicago *Tribune* had been determined to put out a newspaper on the fire and its aftermath. These dedicated employees worked well into the evening, but the printing presses melted just before the building went up in flames.

The Aftermath

By dawn the following day, Chicago still burned. Fortunately, at 8:00 A.M., firefighters arrived from Milwaukee, and were later joined by help from Cincinnati and Dayton, Ohio. Finally, on Monday afternoon, it looked as though the fire was starting to burn itself out. But not until later that night did a much-needed rainfall **extinguish** the last of the flames.

Seventeen thousand homes and buildings were destroyed in the Great Chicago Fire, leaving about 90,000 residents without shelter. The total **monetary** (financial) damage to the city was just under $200 million. Tragically, between 250 and 300 people were killed.

Yet the unbeatable spirit of the city and its people was evident the next morning as residents began rebuilding. Chicago

Out of Control

The Great Chicago Fire raged out of control for two days—from October 8 until October 10.

city officials moved their offices into a church that survived the flames, and additional police were quickly sworn in to stop looters and restore order. Although the Chicago *Tribune*'s printing presses and building had been destroyed, the staff still managed to put out a newspaper the next day. Its front page editorial may have best reflected the city's firm resolve not to give up. It boldly stated "Chicago shall rise again."

After the fire, small temporary homes called shanties were built until more suitable housing became available.

Community Spirit

Although thousands of homes and buildings were lost during the blaze, citizens of Chicago pulled together, and by the end of the first week after the fire, 6,000 temporary structures were built.

MEYERS, CROWN & WALLACH
HIGH STANDARD
CLOTHING

MAURICE BLUM
CLOTHING SPECIALTIES

23-29

HARRIS BROS
MFRS OF
MENS CLOTHING

BERNSTEIN & MEYERS
CLOAKS AND SUITS

THE HATTERS' FUR EXCHANGE 23-29

Firefighters hose down the smoldering fire in the Asch Building, where 146 people died.

Chapter Two

The Triangle Shirtwaist Fire

It was past four o'clock on the afternoon of March 25, 1911, and the nearly six hundred men and women at New York City's Triangle Shirtwaist Company were finishing up the day's work. Many were working overtime to catch up on a backlog of orders. Their job was never easy, and sometimes it was nearly unbearable. These employees worked long hours in a hot, crowded, unsanitary factory. They would often work between 14 and 16 hours a day, and were sometimes paid less than $1.50 a week. Someone who sewed a coat that sold for seventy-five dollars might be paid about seventy-five cents for his or her work.

The company owners were more concerned with profits than with maintaining a safe workplace, so they hired employees with few other job choices. Usually, the workers at companies like the Triangle Shirtwaist Company were immigrants who had recently arrived in the United States from Italy, Germany, Russia, and Hungary. Most spoke little English, and few were over twenty years old. Some were as young as thirteen.

The Asch Building

The Triangle Shirtwaist Company was located on the top three floors of the Asch Building in New York City. At that time it was not uncommon for garment-industry factories to occupy the top floors of buildings, since ground-level space was scarce. The Triangle Shirtwaist Company had **seamstresses** on the eighth and ninth floors, with its executive

Fire Inspections

The International Association of Fire Chiefs (IAFC), founded during the late 1940s, prompted a nationwide effort in the United States to promote home and building inspections. These inspections significantly helped to reduce the risk of fire.

offices on the tenth. The lower floors contained as many sewing machines as could be crammed into the space, providing the employees with very little room.

Built in 1900, the Asch Building was fairly new. It was supposed to have been a modern, **fireproof** structure, but the materials inside the factory were hardly fire resistant. The place was filled with wooden tables and cloth for making shirts. In addition, there were obstacles blocking exits that could turn any fire there into a major disaster.

To reach the elevator under normal circumstances, employees had to walk down a narrow hallway that was lined with boxes of clothing. This forced workers to walk to the elevator single file, allowing company inspectors to search employees, to make sure they hadn't stolen anything while on duty. There were also two staircases leading out of the building but, to further prevent theft, one of the exit doors was bolted shut, and the other door opened only from the outside.

Fire Hazards

The building's only fire escape had not been maintained, and its fire hose was badly damaged. An old, faded NO SMOKING sign hung on the wall, but workers ignored its warning. The owners of the Triangle Shirtwaist Company knew they were violating city fire codes, but like many **"sweatshop"** owners they believed they could get away with it.

Sweatshops

State laws were enacted in the 1880s to regulate conditions in sweatshops. These laws were referred to as the "anti-sweating" laws. Unfortunately, some sweatshops still exist throughout the world today.

21

Many sweatshop owners crammed their workers into unsafe, hazardous rooms similar to the one shown here.

On the day of the fire, the factory was filled with fabric that could readily go up in flames. Baskets of silk and lace lined nearly all the factory's aisles and hallways, while the scrap bins

Fabric and wooden sewing tables fueled the fire as it spread rapidly throughout the Triangle Shirtwaist Company.

brimmed over with thread and cut pieces. In addition, numerous completed garments hung from lines strung above the sewing machines.

A Fight for Life

Late that afternoon, a horrendous fire began in a rag bin on the factory's eighth floor. Seeing the flames from her sewing machine, a young woman screamed, "Fire!" Within seconds, the factory foreman and manager raced to the bin with buckets of water to extinguish it. But the fire had already grown too large. Other men grabbed the building's fire hose to quench the flames, but it fell apart in their hands.

Fueled by the material and finished garments, the fire spread rapidly. Terrified, the women tried to make their way through the crowded aisles to the elevator. But the elevator only held a dozen people, and now hundreds wanted to get on

Many workers tried to make it through this window and down the fire escape to safety. Only a few survived.

it at once. Others who tried to escape down the stairwell, found the exit bolted as usual. A number of women tried to push it open but couldn't. At that point, there was no turning back. The fire had consumed much of the factory floor and was not far behind them. Rosie Safran, a seamstress who survived the fire, later described what it was like. "If we couldn't get out we would be roasted alive," she remembered. "The locked door that blocked us was half of wood; the upper half was thick glass. Some girls were screaming, some were beating the door with their fists, some were trying to tear it open. Someone broke the glass part of the door with something hard and heavy. I climbed or was pulled through the broken glass."

Many people tried escaping through the staircase leading to the door that only opened from the outside, but they perished on the spot. Finally, a group of men managed to pull the door open and a handful of women escaped. Rather than be burned alive, others jumped out of the windows, only to die from the fall. A number of girls stood on the window ledge hoping to climb down the fire company's ladders. But when they realized that the ladders only reached as far as the floor below them, they jumped instead. Tragically, they joined the pile of corpses that had begun to fill the street below.

Plan of Action

When the chief officer arrives at the scene of a fire, he designates orders to his department based on a standard plan of action. The firefighters must:

1. Find the precise location of the fire.
2. Rescue any endangered occupants of the building.
3. Confine the fire to the area it controls.
4. Attack and extinguish the flames.
5. Search out and extinguish all hidden flames before declaring the fire out.

Coffins are brought in for the corpses that line the sidewalks around the factory.

With little time to spare, firefighters tried other means to save those trapped inside the building. "One girl jumped into a horse blanket held by firemen and policemen," a writer for the New York *Times* newspaper reported. "The blanket ripped like a cheesecloth and her body was mangled beyond recognition."

Before the fire disabled the elevator, elevator operator Joseph Zito made a number of trips to bring young women out of the building. But he never reached those waiting for him on the ninth floor. Each time the elevator arrived at the eighth, it was mobbed by workers trying to leave. In a desperate attempt to save themselves, some of the women on the floor above jumped down the elevator shaft. Tragically, they died as their bodies smashed against the top of the descending elevator.

Company owners Max Blanck and Issac Harris safely left the fiery **inferno** that the factory had become. As it happened, Blanck's children, along with their governess, had come to visit him at work that day when the fire broke out. The group went up to the building's roof and managed to escape to an adjacent building.

But others experienced a more gruesome fate. One hundred and forty-six people died in the Triangle Shirtwaist fire, which actually lasted less than twenty minutes. Yet its horrific

Statistics
America's fire death rate is one of the highest in the industrialized world. Fire kills more than 4,000 people and injures more than 27,000 individuals each year. Approximately 100 of these fatalities are firefighters who have died in the line of duty.

Family members line up, waiting to identify loved ones who were killed in the fire.

27

impact cannot be underestimated. According to the newspaper *The World*, New York City's fire chief, who was "a man used to viewing horrors [came] back down . . . [from the building] with quivering lips. In the drifting smoke he had seen bodies burned to bare bones, skeletons bending over sewing machines."

An Angry Public

The public's response to the tragedy was powerfully emotional as well. Shocked by the lack of safety **precautions** at the company, more than a thousand people attended a funeral mass for the fire's victims. Then in early April, eighty thousand individuals joined a protest march in which they followed an empty horse-drawn hearse down the street to draw attention to the problems of factory workers.

But the injustice continued when, in December of that year, a grand jury cleared the factory owners of **manslaughter** charges. The cause of the fire was never established. The jurors determined that the exit door had probably been mistakenly bolted by one of the workers. Those outraged by the decision stressed that the jury was all male, while most of the fire victims had been females. Before long, the owners were back in business.

In 1914, three years after the fire, civil suits were brought against the company owners by relatives of twenty-three victims. Each family was awarded seventy-five dollars for the loss of their loved one.

Federal Fire Prevention

In 1974, the Federal Fire Prevention and Control Act was signed into law. From this law the United States Fire Administration (USFA) and the National Fire Academy (NFA) were created. Through public education, data collection, research, and training efforts, these organizations have helped reduce fire deaths, making communities and citizens safer.

The workers' deaths were not totally in vain. The fire made a formerly indifferent public aware of the inhumane and dangerous conditions in numerous workplaces. As a result of the Triangle Shirtwaist tragedy, many fire-prevention measures have become standard in businesses throughout the United States.

Although the contents of the Triangle Shirtwaist Company were consumed by the fire, the walls, floors, and ceilings remained intact—proving that the Asch Building was fireproof.

The charred remains of the Cocoanut Grove marquee

Chapter Three

The Cocoanut Grove Nightclub Fire

The Cocoanut Grove may sound like a group of palm trees in the tropics, but in the early 1940s it was a large, lavishly decorated Boston nightclub. This popular night spot had a spacious dance floor and dining area. It offered extravagant stage shows, magic routines, and big-band dance music. There were also other rooms—quiet, dimly lit areas with just a piano player and bar for customers who only wanted to talk and unwind.

On most weekends, the Cocoanut Grove was filled with partygoers out for a good time. But on Saturday, November 28, 1942, it was especially crowded. A victory party for the Boston College football team was to be held there that night, and even though the team had unexpectedly lost earlier in the day, many people decided to enjoy themselves at the club anyway. This included some well-known personalities, (such as the cowboy movie star Buck Jones) and a small group of housewives who had saved for months to treat their husbands to the afternoon football game and a night out at the Cocoanut Grove.

"Good Condition"

While the furnishings might have been somewhat flammable, the Cocoanut Grove's tropical interior undeniably added to the club's atmosphere. Behind the revolving doors at the main entrance, the club looked like an island paradise with blue satin skies and flowers, palms, and coconuts made of paper, cloth, and bamboo. To section off different parts of the establishment, a number of leather-covered artificial walls were also installed.

One week before the fire, Lieutenant Frank Linney of the Boston Fire Department had inspected the building, describing the premises as being in "good condition." The fire inspector's report further indicated that there were both "a

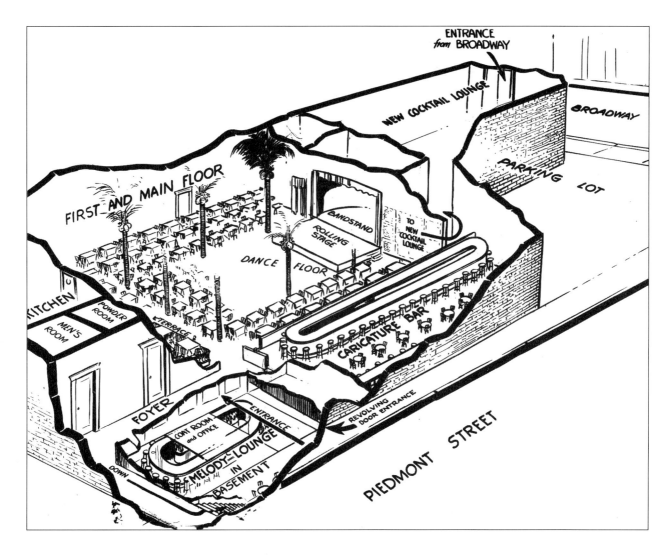

sufficient number of exits" and a "sufficient number of **fire extinguishers**."

But that Saturday night, conditions at the Cocoanut Grove were hardly ideal for dealing with a fire. While the nightclub was designed to hold from four hundred to six hundred people, more than one thousand patrons had made their way inside the club. So as not to lose business or irritate customers,

This floor plan shows the layout of the rooms in the Cocoanut Grove Nightclub.

the waiters were instructed to set up additional tables on the dance floor and in other parts of the club. This may have sat- isfied patrons waiting to get in, but it also blocked aisles, pro- hibiting large numbers of people from hurriedly exiting the building in an emergency.

Stanley Tomaszewski is sworn in before testifying at his hearing.

Stanley Tomaszewski

Ironically, the person who started the fire at the Cocoanut Grove should never have been there. Sixteen-year-old high school honor student and football player Stanley Tomaszewski was employed at the club as a bar boy, although he was too young to be legally working in a bar. Tomaszewski was helping the bar- tenders that night in the club's basement Melody Lounge.

The lounge was lit by electric bulbs concealed inside coconuts on the artifi- cial palm trees. Just after ten o'clock, a

34

bartender asked Tomaszewski to replace a lightbulb in one of the coconuts. The boy did so, but as the lounge was fairly dark he lit a match to see the light socket. He blew out the match, never realizing that the tiny flame had already ignited one of the palms near the ceiling.

Within seconds, a fire had begun spreading rapidly. The bartender spotted the flames, and he and Tomaszewski tried to pull down the draperies, which had already turned into a fiery curtain. But it was no use—a large portion of the lounge was soon ablaze, and the fire showed no signs of stopping. Terrified, Tomaszewski fled from the building through a door in the kitchen.

The Victims

Within minutes, a thick cloud of smoke had filled the room, making it even more difficult for the confused and frightened partygoers to leave. About 130 customers headed toward the single narrow staircase, tripping over one another in their panic. The flames reached the stairs before many could escape.

Meanwhile, people in the main dining room above the lounge were still unaware of the fire. The floor show was about to begin when a young woman, who had made it upstairs from the Melody Lounge, ran across the dance floor with her hair on fire. Before seated patrons realized the danger, fire **engulfed** the dining room, killing all the guests in the main dining room.

Causes of Fire

Smoking and the careless use of matches cause the greatest number of fires in the United States.

Fire inspectors sift through the remains of the nightclub.

Now fearing for their lives, patrons in other areas of the nightclub tried to leave through the entrance's revolving doors. But the stampeding herd of people jammed the doors after a few got out. Although there was an emergency exit next to the revolving doors, for some unknown reason that door

Ventilation

Ventilation is an important factor in fighting a fire. To ventilate a building, firemen will immediately break windows, open vents, or cut a large hole in the roof of a house or building. These openings permit the pent-up heat, smoke, and toxic gases to escape, preventing a possible explosion from occurring.

was bolted shut. People in the newer section of the club known as "The New Cocktail Lounge" found themselves trapped as well.

The Heroes

Some people in the club remained calm and made their way down to the basement kitchen, hoping to find an exit. A waiter named Henry W. Bimler also went to the basement to leave through the employees' entrance, but it was locked. On his way back up, he was stopped by two young ladies who begged him to help them. Bimler took the women back to the kitchen and led them into the Cocoanut Grove's walk-in refrigerator. Though somewhat chilled, the small group remained safe in the spacious cooler until they were later rescued by firemen.

With his hands badly burned, Mickey Alpert (center) is escorted from the scene after rescuing others trapped inside the club.

Others on the nightclub's staff acted heroically as well. After escaping the fire himself, bandleader Mickey Alpert returned to the building several times to rescue others. As a result, he was badly burned, but he eventually recovered. Another hero that night was a nineteen-year-old dancer in the show named Marshall Cook. When fire and smoke blocked his way to the door, the young man broke a window in the second-floor dressing room and helped more than

The Red Cross

Clara Barton was the first person to organize a Red Cross chapter in the United States after working with the thousands of wounded soldiers in the Civil War. In 1881, the American Red Cross was founded to serve America in peace and in war, during times of disaster and great distress. Today, the actions of the organization are guided by humanity, impartiality, neutrality, independence, voluntary service, unity, and universality.

thirty-five show people and stage hands to safety on the roof of the next building.

One passerby who also acted admirably was Joseph Lawrence Lord, a naval officer and former fireman, who spotted the fire while driving by in his car. Stopping to help, he broke a window in the flaming nightclub and raced in. "The smoke was choking and thick," he recalled, "but no different really than any fireman meets during many a fire. I crawled along on my hands and knees and then bumped into . . . three women and two men, and one by one, I dragged them to the window I had jumped through. I hoisted them up to the sill and then yelled. . . . Firefighters gave me a hand and pulled them through the window to safety."

Others helped as well. Many of the men who made it out of the building stayed to assist the firefighters. A priest arrived to administer last rites to the dying, while the Red Cross sent trained first-aid volunteers and disaster-relief workers.

Nina Underwood, a nurse's aide with the Red Cross, knew what was needed as soon as she arrived on the scene. "I had to

knock down two survivors with jujitsu," she noted, "because they got out of hand. I questioned a girl who was pretty high on liquor and she said she didn't know there was any trouble until the smoke grew heavy. . . . I gave artificial respiration three times . . . and shock treatment once."

More than a hundred ambulances from surrounding areas came to transport the injured. When even these weren't enough, almost as many taxicabs were used as well. A long line of trucks helped haul away the dead.

Victims are passed above the crowd to ambulances waiting to transfer the injured and the dead.

Safety Regulations

Four hundred and ninety-one people died in the Cocoanut Grove fire. Although young Stanley Tomaszewski received a good deal of publicity for starting the blaze, the nightclub and the city of Boston itself were also at fault. While the fire inspector reported that the club was in good condition, its flammable decorations and inexpensive wiring had actually set the stage for disaster.

At the time, Boston had very few fire **regulations** to ensure the safety of those enjoying its nightlife. Robert Moulton of the National Fire Protection Association commented, "Nightclubs are commonly located in

Barnett Welansky (middle), owner of the Cocoanut Grove Night-club, was convicted of manslaughter on April 15, 1943.

NFPA

The National Fire Protection Association (NFPA) was founded in 1896. The NFPA produces the National Fire Codes—275 rules and standards covering all areas of fire safety. The fire codes are used in nearly every country in the world. Today, most buildings in the United States are required to meet the standards developed through this system. This is called being "up to code."

old buildings . . . and practically every known rule of fire safety is violated."

As a result of the Cocoanut Grove incident, the fire safety regulations imposed on theaters were extended to nightclubs. Fireproof sprinkler systems and clearly marked exit signs were required, and room-capacity limits were enforced. It was hoped these measures would help ensure that the Cocoanut Grove tragedy would never be repeated.

A funeral mass is said for four brothers who died in the Cocoanut Grove fire.

Spectators flee the burning circus tent.

Chapter Four

The Hartford Circus Fire

It was a hot, humid day in July 1944 in Hartford, Connecticut, but many of the young people weren't thinking about the weather. Instead, they were excited about the circus being in town. The Ringling Brothers and Barnum & Bailey Circus billed itself as "the greatest show on earth," and that afternoon more than six thousand people eagerly entered the huge tent known as the Big Top to see the show.

Workers prepare to unroll a canvas tent similar to the one used in the Ringling Brothers and Barnum & Bailey Circus.

The circus offered entertainment at a stressful time for the country. The United States had entered World War II, and there were few people whose lives hadn't been affected by the war in some way. With valued materials reserved for the war effort, the circus had to use whatever supplies were available. Since **canvas** was now primarily used for military purposes, its nearly 200-foot (60 m)-wide canvas tent had to be patched to last longer than in prewar times. Circus owners tried to waterproof and preserve the tent by treating it with a mix of **paraffin** and gasoline, but the process also made the tent dangerously flammable.

Wartime shortages also resulted in other circus hazards. In the past, the show had always used thick, strong fire-resistant **hemp** ropes for aerial acts, as well as to hoist the tent. But during the war, the high-grade rope wasn't available, forcing the circus to use a less fire-resistant type.

Thursday, July 6, 1944

The show began at about two o'clock in the afternoon. The twenty-nine-piece orchestra played a number of tunes. Then came the grand march of the elephants, followed by clowns, acrobats, and performing lions and tigers. It was during the high wire act, the Flying Wallendas, when circus bandleader Merle Evans noticed a small, flickering flame in the canvas tent.

Hoping to call for help without causing a panic, Evans quickly had his musicians play the stirring march "Stars and Stripes Forever." In the circus world, an abrupt musical change to a lively tune was a signal for help. The response was immediate. The Fire Department was called and tent ushers, along with other

The Greatest Show on Earth

The Ringling Brothers and Barnum & Bailey Circus has been a part of American tradition for more than a century. Bought by the Ringling Brothers—Albert, Otto, Alfred, Charles, and John—in 1907 from Barnum & Bailey, it was the largest traveling circus in the world.

circus staff, came running with buckets of water to douse the flames.

At first, the audience did not panic. Those near the area thought it was just a small fire that would be put out easily. People simply proceeded in an orderly manner to the nearest exit believing that they'd soon return to their seats for the rest of the show. The bandleader did his best to keep things calm by continuing the music as if there was nothing to be concerned about.

But no one imagined the speed at which the blaze would spread. The fire raced up and across the canvas as a stiff breeze

fanned the flames still higher. Flaming ropes plummeted and the large tent's center poles began to fall.

Within minutes, a full-scale panic broke out as thousands of people realized the seriousness of the situation and attempted to flee. But unfortunately many were prevented from leaving.

The circus tent was designed with 4-foot (1.2 m)-high steel passageways positioned on both sides of the reserved-seat section. This was for the audience's protection, since the circus's wild animals walked from their cages to the show ring. These seats gave the audience, sandwiched in the middle, an excellent view of the show. But when the fire broke out, they had to leap or climb over the passageways to exit the tent.

Canvas Tents

Circus and carnival shows were first presented beneath a canvas tent in 1826. Before then, performers worked on an outdoor stage.

47

Lives Lost

If the blaze had started just minutes later, circus employees would already have removed the portable animal runways. Although some people scrambled over the runways, many of the young children and older adults found it difficult. In the chaos, a number of people fell, and the crowd stepped over and on top of them. A young boy pleaded for someone to help him lift his grandmother, who had fallen; he was ignored.

Thomas C. Murphy, a reporter for the Hartford *Courant*, had brought his five-year-old son to the show. Murphy didn't want to attend the performance that day because of the extreme heat, but his young son persuaded him to go. When the fire broke out, Murphy tried to make his way through the crowd in front of the animal runways. "I was slammed against the steel barrier and my knee caught momentarily between the bars," he later recounted. "Then, taking my five-year-old son in my hands, I tossed him over the barrier to the ground beyond. The flames

Firefighters hose down the burning animal cages that prevented some spectators from leaving the fiery tent.

at this point were nearly overhead, and the heat was becoming nearly unbearable."

Outside, a woman who had been separated from her children tried to go back into the tent but was stopped by police. Even though the fire department arrived promptly, the fire had spread too rapidly for them to do very much. Firefighters pulled out any victims still showing signs of life and hosed down the charred circus ruins. The loss of life and property damage had been considerable. One hundred and sixty eight people were killed, and about five hundred others were injured.

A fleet of ambulances were needed to take the injured to various hospitals. Other vehicles took unidentified dead bodies to the state armory until family members could claim them.

Firefighters

Aboard a **pumper** or a ladder truck, firefighters are expected to leave the fire station within 30 to 45 seconds after an alarm has been received. A firefighter must be physically fit and adaptable to teamwork.

The Connecticut state armory housed the dead until family members could identify them.

More than two-thirds of those killed that day were young children.

One especially pretty six-year-old girl who died in the fire was never identified, although her face was unmarred by the flames. She was referred to by rescuers as "Little Miss 1565" (referring to her body-tag number at the temporary **morgue)**. For years thereafter on the anniversary of the fire, Hartford police put flowers on her grave. "It just doesn't seem possible," an officer noted, "that a child like that little one could have disappeared from her own small world without somebody noticing that she had gone and never came back." Yet it happened.

The Investigation

An official investigation was held after the fire. Five circus officials were charged with technical manslaughter, and other serious charges were leveled against the circus management. Hartford County State Attorney H. M. Alcorn, Jr., noted that there had been " . . . inadequate fire-fighting equipment on the grounds . . . and inadequate personnel to operate the small amount of equipment available." Other fire-code violations were identified as well. The Hartford tragedy even affected government policy. Within one day of the blaze, fireproof tent material formerly reserved for the military was made available to circuses throughout the United States.

There were many theories about how the fire started, but no one knew the whole truth until six years after the blaze. In

July 1950, twenty-year-old Robert Dale Segee confessed to purposely setting the fire when he was just fourteen years old. Segee, who had a difficult childhood, had often been in trouble with the law.

New Precautions

After the fire, the Ringling Brothers and Barnum & Bailey Circus permanently changed its policies. It would never again use traditional canvas circus tents. Instead, performances would only be held in indoor stadiums, or in large open areas such as ballparks.

Six victims of the Hartford Circus fire are laid to rest.

The circus still faced a number of lawsuits from the families of those who died. None of the relatives were forced to go to court to collect what was owed to them—instead, dollar amounts were determined by a team of impartial judges. The total figure due to the families was $4 million, but the circus's insurance company would cover only $500,000. Therefore, much of the show's profits over the next decade were used for these settlements, which have frequently been described as being among the fairest in modern times.

Timeline

1752	The first fire insurance company is established in America, called the Philadelphia Contributionship, which remains in existence today
1840	Dr. W. F. Channing invents the fire-alarm telegraph system
1845	A fire breaks out in a theater in Canton, China, killing more than 1,670 people
1871	The Great Chicago Fire destroys 17,000 buildings and kills between 250 and 300 people
1871	On the same day as the Great Chicago Fire, October 8, 1871, a forest fire breaks out in Peshtigo, Wisconsin, killing 1,182 people
1871	New York introduces an automatic telegraph alarm system
1873	The International Association of Fire Chiefs is organized in Baltimore, Maryland
1889	A committee on fire departments, fire patrols, and water supply is established, called the National Board of Fire Underwriters
1896	The National Fire Protection Association (NFPA) is established
1911	The Triangle Shirtwaist factory fire kills 146 employees
1930	A fire breaks out in a prison in Columbus, Ohio, killing 320 prisoners
1940	The International Association of Fire Chiefs (IAFC) sets standards for sprinklers, fire doors, extinguishers, and other equipment

1942	A fire is accidentally set at the Cocoanut Grove Club in Boston, killing 491 patrons
1944	A tent fire at the Ringling Brothers and Barnum & Bailey Circus in Hartford, Connecticut, kills 168 people
1968	The Fire Research and Safety Act is passed by Congress
1968	The National Commission on Fire Prevention and Control is established by the Fire Research and Safety Act
1973	A report called *America Burning* is issued to President Richard Nixon by the National Commission on Fire Prevention, focusing on America's fire problem
1974	The Federal Fire Prevention and Control Act is signed into law, creating the United States Fire Administration (USFA) and the National Fire Academy (NFA)
1985	Fifty-three fans are killed when a fire erupts at a soccer game in Bradford, England
1990	The Hotel and Motel Fire Safety Act is passed into law by Congress to promote fire and life safety in hotels, motels, and other places of public accommodation
1990	Eighty-seven people die in a fire at the Happy Land Social Club in the Bronx, New York
1992	The United States Congress creates the National Fallen Firefighters Foundation to honor American firefighters who died in the line of duty
1994	A toy factory fire kills 213 in Bangkok, Thailand
1996	The National Arson Prevention Initiative (NAPI) is announced by President Bill Clinton in an effort to raise the public's awareness about how arson fires can be prevented

Glossary

canvas—a coarse, strong cloth used for tents, sails, and clothing

combustible—something that burns easily

dispatcher—someone who communicates an important message to another

engulfed—covered or swallowed up

extinguish—to put out a flame, a fire, or a light

fire extinguisher—a metal container with chemicals and water inside it that can be used to put out a fire

fireproof—something made from material that will not burn easily

flammable—something that is easily set on fire and burns rapidly

hemp—a strong, tough plant fiber sometimes used to make rope

ignite—to set on fire

inferno—a fiery place

inquiry—an investigation

manslaughter—the crime of killing someone without intending to do so

monetary—having to do with money

morgue—a place where dead bodies are kept until claimed by a relative for burial

paraffin—a waxy, white, water-repellent substance

precaution—something you do in order to prevent something dangerous or unpleasant from happening

pumper—a fire truck equipped with a pump

regulations—official rules or orders; the act of controlling or adjusting something

reinforcement—something that strengthens

seamstresses—women who sew for a living

sweatshop—a shop or factory in which workers are employed for long hours at low wages and under unhealthy conditions

ventilation—a means of providing fresh air

To Find Out More

Books

Bortz, Alfred B. *Catastrophe! Great Engineering Failure—and Success*. New York: Scientific American Books for Young Readers, 1995.

Cone, Patrick. *Wildfire*. Minneapolis: Carolrhoda Books, 1997.

Masoff, Joy. *Fire!* New York: Scholastic, Inc., 1998.

Ready, Dee. *Fire Fighters*. Danbury, CT: Children's Press, 1997.

Sherrow, Victoria. *Triangle Factory Fire*. Brookfield, CT: Millbrook Press, 1995.

Organizations and Online Sites

American Red Cross
Attn: Public Inquiry Office
1621 N. Kent Street
Arlington, VA 22209
http://www.redcross.org/
This organization has provided assistance to victims of disasters for more than a decade.

The Great Chicago Fire and the Web of Memory
http://www.chicagohs.org/fire/intro
Maintained by the Chicago Historical Society and Northwestern University, this site marks the 125th anniversary of the Great Chicago Fire.

The Great Triangle Fire/Discovery Channel Online
http://www.discovery.com/DCO/doc/1012/world/history/triangle-fire/triangle1.html
This site provides a word-by-word account of one of the most horrific factory fires in history.

Smokey the Bear Home Page
http://www.smokeybear.com/
Loaded with fun games and activities for kids, this page explains how forest fires can start, and what we can do to prevent them.

Sparky's Firefighting Page

http://www.wenet.net/sandd/sparky.htm

This fact-filled page provides links to more than fifty fire-related websites, including exit drills kids can practice in their homes in case of a fire.

USFA Kids Page

http://www.usfa.fema.gov/kids/index.htm

A website designed by the United States Fire Administration that provides helpful tips for kids, teachers, and parents, that can help save a life during a fire.

A Note on Sources

In writing on disastrous fires, I consulted a number of sources. These included the following books: *Mrs. O'Leary's Comet! Cosmic Causes of the Great Chicago Fire*, by Mel Waskin; *Great Fires of America*, by the Editors of *Country Beautiful*; *Inferno! Fourteen Fiery Tragedies of Our Time*, by Hal Butler; *Man Made Catastrophes*, by Lee Davis; *Eyewitness To Disaster*, by Dan Perkes; *From the Pages of* The New York Times, edited by Arleen Keylin and Gene Brown; and the *Triangle Fire*, by Leo Stein.

The magazines and newspapers I consulted included The New York *Times*, *American Historical Review*, and the *Interdisciplinary History*. Further valuable information was provided for the book by the Chicago Historical Society, The National Fire Protection Association, and the Metro Dade Fire Department of Miami, Florida.

—*Elaine Landau*

Index

Numbers in *italics* indicate illustrations.

Alcorn, H. M., Jr., 50
Alpert, Mickey, 37, *37*
American Red Cross, 38
Asch Building, *18*, 20–21, *29*

Bales, Richard F., 11–12
Barton, Clara, 38
Bimler, Henry W., 37
Blanck, Max, 27
Boston College, 32
Boston, Massachusetts, 31–41

Channing, W. F., 12
Chicago, Illinois, 7–17. *See also* Great Chicago fire
Chicago Title Insurance Company, 11
Chicago *Tribune*, 16, 17
Cincinnati, Ohio, 15, 16

Cocoanut Grove Nightclub fire, *30*, 31–41, *33*, *34*, *36*, *37*, *39*, *40*, *41*
Cook, Marshall, 37–38

Dayton, Ohio, 16

Evans, Merle, 45, 46

Federal Fire Prevention and Control Act, 28
Fire-alarm telegraph system, 12
Fire death rates, 27
Fire escapes, 21, *24*
Fire extinguishers, 24, 33, 34
Firefighters, 9, 12, 13, *13*, 14, 15, 16, *18*, 25, 26, 27, 38, *48*, 49

Fire inspections, 20, 32–33,
    36, 40
Fire insurance, 8, 11, 51
Fire regulations, 28, 40–41,
    50
Fire sprinkler systems, 41
Fire watch, 12–13
Flying Wallendas, 45

Gasoline, 34, 44
Goodspeed, E. J., 14–15
Great Chicago fire, 7–17, *7,
    8, 9, 11, 13, 14, 15, 17*

Harris, Issac, 27
Hartford Circus fire, *42,*
    43–51, *44, 45, 46, 47,
    48, 49, 51*
Hartford, Connecticut,
    43–51
Hartford *Courant*, 48
Hemp ropes, 45

International Association of
    Fire Chiefs (IAFC), 20

Jones, Buck, 32

Linney, Frank, 32
"Little Miss 1565," 50

Lloyds of London, 8
Lord, Joseph Lawrence,
    38

Milwaukee, Wisconsin, 15,
    16
Moulton, Robert, 40–41
Murphy, Thomas C. 48–49

National Fire Academy
    (NFA), 28
National Fire Codes, 40
National Fire Protection
    Association (NFPA), 40
New York City, 19–29
New York *Times*, 26

O'Leary, Katie, 10–11, *11,*
    12

Paraffin, 44

Red Cross, 38
Ringling Brothers and
    Barnum & Bailey
    Circus, 43, *44, 45,* 51.
    See also Hartford Circus
    fire

Safran, Rosie, 25

St. Louis, Missouri, 15
Segee, Robert Dale, 51
Smoking, 21, 35
Sullivan, Daniel (Peg Leg),
    11–12
Sweatshops, 21, *22*

Tents, 44, *44*, 47
Tomaszewski, Stanley,
    34–35, *34*, 40
Toxic gases, 36
Triangle Shirtwaist
    Company, 20, 21, *23*,
    28, *29*

Triangle Shirtwaist fire, *18*,
    19–29, *23*, *24*, *26*, *27*, *29*

Underwood, Nina, 38–39
United States Fire Adminis-
    tration (USFA), 28

Ventilation, 36

Welansky, Barnett, *40*
Williams, Robert, 9, 14
World, The, 28

Zito, Joseph, 26

About the Author

Popular author Elaine Landau worked as a newspaper reporter, editor, and as a youth services librarian before becoming a full-time writer. She has written more than one hundred nonfiction books for young people. Included among her many books for Franklin Watts are the other Watts Library titles on disasters: *Air Crashes, Maritime Disasters,* and *Space Disasters.* Ms. Landau, who has a bachelor's degree in English and journalism from New York University and a master's degree in library and information science from Pratt Institute, lives in Miami, Florida, with her husband and son.